Under the volcano

Claire Llewellyn

Kaeden Corporation
PO Box 16190
Rocky River, Ohio 44116
1-800-890-READ (7323)
www.kaeden.com

Under the Volcano was originally published in the United Kingdom in 2005. This edition is published by arrangement with Oxford University Press.

Acknowledgements

The publisher would like to thank the following for permission to reproduce photographs: p4/5 Travel Ink Photo Library; p15 Corbis/Ressmeyer; p17 AIPIX/Superbld; p18 Corbis/Bettmann; p19t Corbis/Richard T Nowitz, b Corbis/Roger Ressmeyer; p20t Scala Art Resource/Ministero Beni e Att. Culturali, b Scala Art Resource/Photo Scala/Florence/Fotografica Foglia; p21t Alamy images, c Scala Art Resource/Photo Scala, Florence, b Corbis/Mimmo Jodice; p22/23 Geoscience Features; p23t Art Directors & Trip Photo Library/Photographers Direct; p24t Mary Evans Picture Library, b Corbis/alinari Archives; p25 Brenton West/Photographers Direct; p26t Alamy/Cephas Picture Library, b Andrew Bain Photography/Photographers Direct; p27t Corbis/Mimmo Jodice, b Alamy/Photos 3; p28 Corbis/Reuters; p29 Geoscience Features

Cover photograph: Corbis/Bettmann

Illustrations by: Julian Baker: p5, p6t, p7, p9, p12/13, p14, p15, p16, p31;
Martin Cottam: p6b, p8, p10, p11, p19

ISBN 978-1-61181-523-8 paperback
Printed in China
CHK -1/15/2016

Contents

The Bay of Naples

The Bay of Naples is a large beautiful bay on the west coast of Italy. On its shores, overlooking the Mediterranean Sea, lies the port of Naples. The city is home to over two million people, and its **suburbs** have merged with nearby towns. Here, people earn their living from trade or farming, producing grains, olives, fruit and vegetables that ripen in the hot summer sun.

Red-hot fact

Mount Vesuvius is the only **active volcano** on the mainland of Europe.

The great peak of Mount Vesuvius dominates the region, like a slumbering giant. At 1281 meters high, it overlooks the farms, the homes, the city and the bay. But this peaceful giant is really a deadly danger: Mount Vesuvius is an active volcano. In the past, its eruptions have brought death and devastation. And, for the millions who live within its reach, it may bring death and devastation again.

Vesuvius erupts

About 2000 years ago, in AD 79, the area around Naples was inhabited by the Romans. The small seaside town of Herculaneum (known today as Ercolano) and the country town of Pompeii must have been very pleasant places to live. Pompeii, particularly, was a bustling place with marketplaces, shops, theaters and baths. Both towns enjoyed views of the green mountain slopes and the blue waters of the bay.

Mount Vesuvius had not erupted for hundreds of years and so no one was aware of its power. But in AD 62 a violent **earthquake** had shaken the region, and many buildings had been damaged. Though the people didn't know it, the earthquake was a warning, and much worse was to come.

In the early afternoon of 24 August AD 79, Mount Vesuvius suddenly roared, and a column of ash and **pumice** was hurled into the sky. The people watched it form a dark umbrella-shaped cloud. Did they sense the horror that was about to begin?

Then and now

In AD 79
No one knew Vesuvius was a volcano.

Today
Vesuvius is one of the most closely studied volcanoes in the world.

Timetable of an eruption

AD 79:

August 24

1:30 p.m.

Eruption begins.

Vesuvius shoots out a 12.5 mile column of ash and pumice.

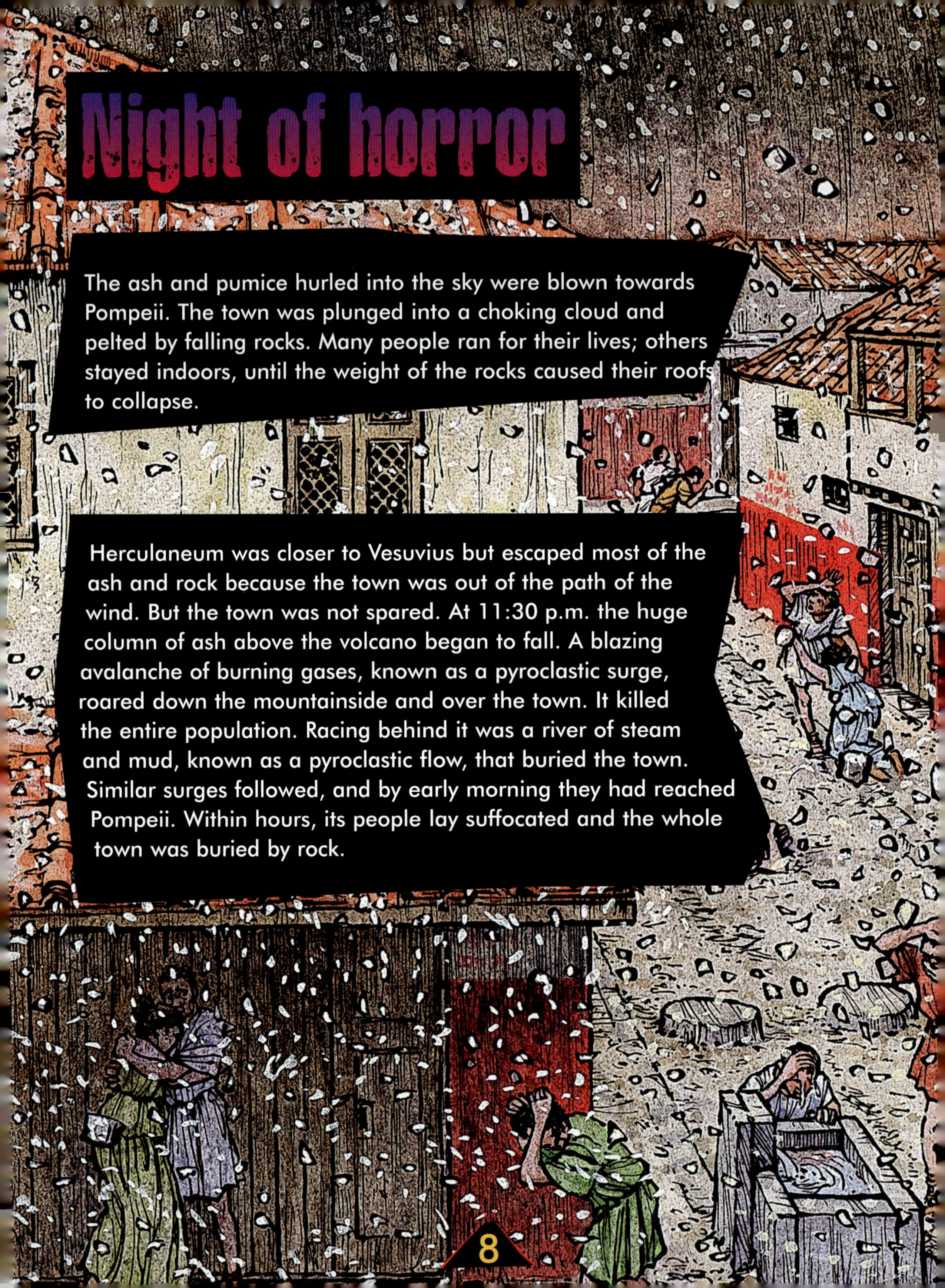

Night of horror

The ash and pumice hurled into the sky were blown towards Pompeii. The town was plunged into a choking cloud and pelted by falling rocks. Many people ran for their lives; others stayed indoors, until the weight of the rocks caused their roofs to collapse.

Herculaneum was closer to Vesuvius but escaped most of the ash and rock because the town was out of the path of the wind. But the town was not spared. At 11:30 p.m. the huge column of ash above the volcano began to fall. A blazing avalanche of burning gases, known as a pyroclastic surge, roared down the mountainside and over the town. It killed the entire population. Racing behind it was a river of steam and mud, known as a pyroclastic flow, that buried the town. Similar surges followed, and by early morning they had reached Pompeii. Within hours, its people lay suffocated and the whole town was buried by rock.

Timetable of an eruption AD 79:

August 24

1:30 p.m. onwards
Ash and pumice fall on Pompeii.

11:30 p.m.
The volcanic column collapses and a pyroclastic surge blasts its way through Herculaneum, killing everyone.

August 25

12:30 – 5:30 a.m.
Further surges bury Herculaneum and threaten Pompeii.

6:30 a.m.
Pyroclastic surge reaches Pompeii, killing everyone.

7:00 – 8:30 a.m.
Pyroclastic flows bury Pompeii.

What is a pyroclastic surge?

In violent volcanic eruptions, a huge column of gas and ash shoots into the air. As it crumbles, burning clouds of ash and gas roar down the hillside at up to 500 km/h. Pyroclastic surges are impossible to escape.

An eyewitness account

How do we know about a volcanic eruption that happened so long ago? In AD 79 an 18-year-old eyewitness called Pliny the Younger was staying at Misenum, on the northern side of the Bay of Naples. His uncle, Pliny the Elder, was commander of the Roman Fleet and sailed across the bay on that fateful day. Years later, his nephew wrote letters to an historian, explaining what had happened.

He steered bravely for the danger zone that everyone else was leaving... The ash already falling, became hotter and thicker as the ships approached the coast... Instead of pushing back as the Captain advised, he decided to push on to Stabiae across the bay, and this is where they landed.

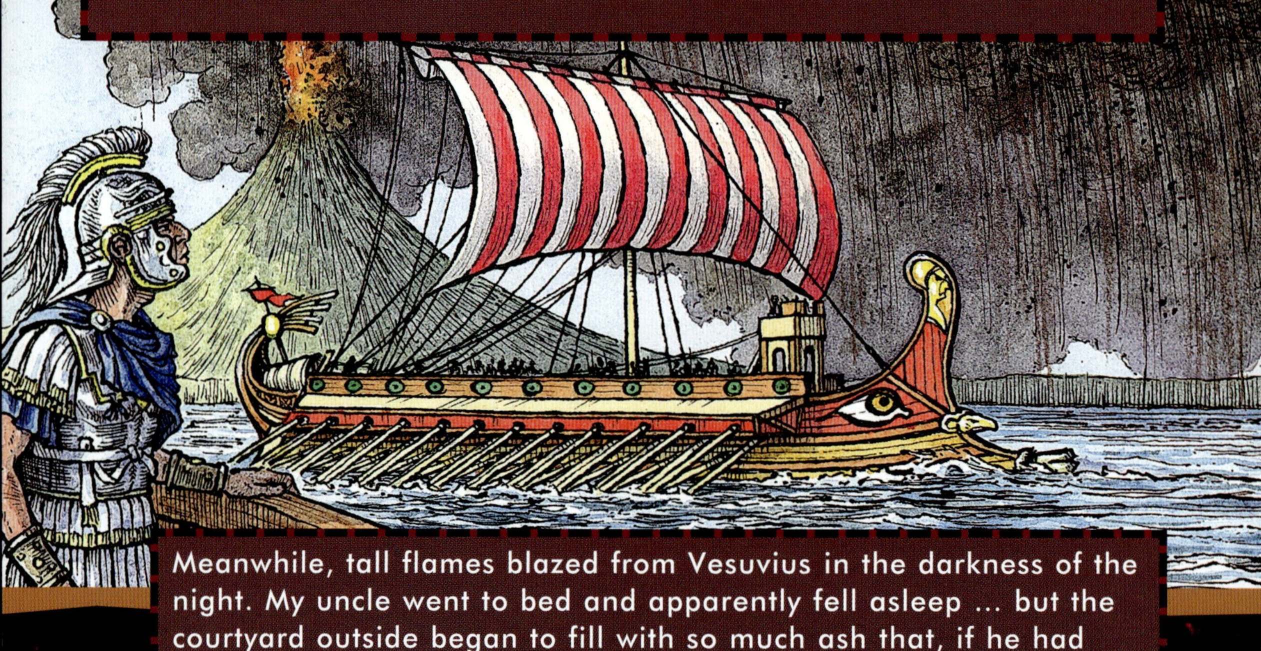

Meanwhile, tall flames blazed from Vesuvius in the darkness of the night. My uncle went to bed and apparently fell asleep ... but the courtyard outside began to fill with so much ash that, if he had stayed in his room, he would never have been able to get out.

They wondered whether to stay indoors or go out into the open, because the buildings were now shaking with more violent **tremors** ... They chose the open country and tied pillows with cloths over their heads for protection.

It was daylight by this time, but they were still enveloped in a darkness that was blacker than any night ... My uncle went to see if there was any chance of escape by sea, but the waves were far too high ... Then, suddenly, flames and the strong smell of sulphur forced the others to flee. He suddenly collapsed and died, because, I imagine, he was suffocated when the dense **fumes** choked him.

Pliny the Younger's letters give a clear description of the eruption. They have helped scientists to understand what happened on the day.

Volcanic eruptions take several forms. The most violent eruptions are known as Plinian eruptions. They are named after Pliny the Elder and his nephew, Pliny the Younger.

Key features of Plinian eruptions:

- huge column of gas
- enormous explosions
- repeated pyroclastic surges
- ash and pumice fall over a wide area
- devastating effects
- thick lava produced only in final stages.

The eruption of Vesuvius:

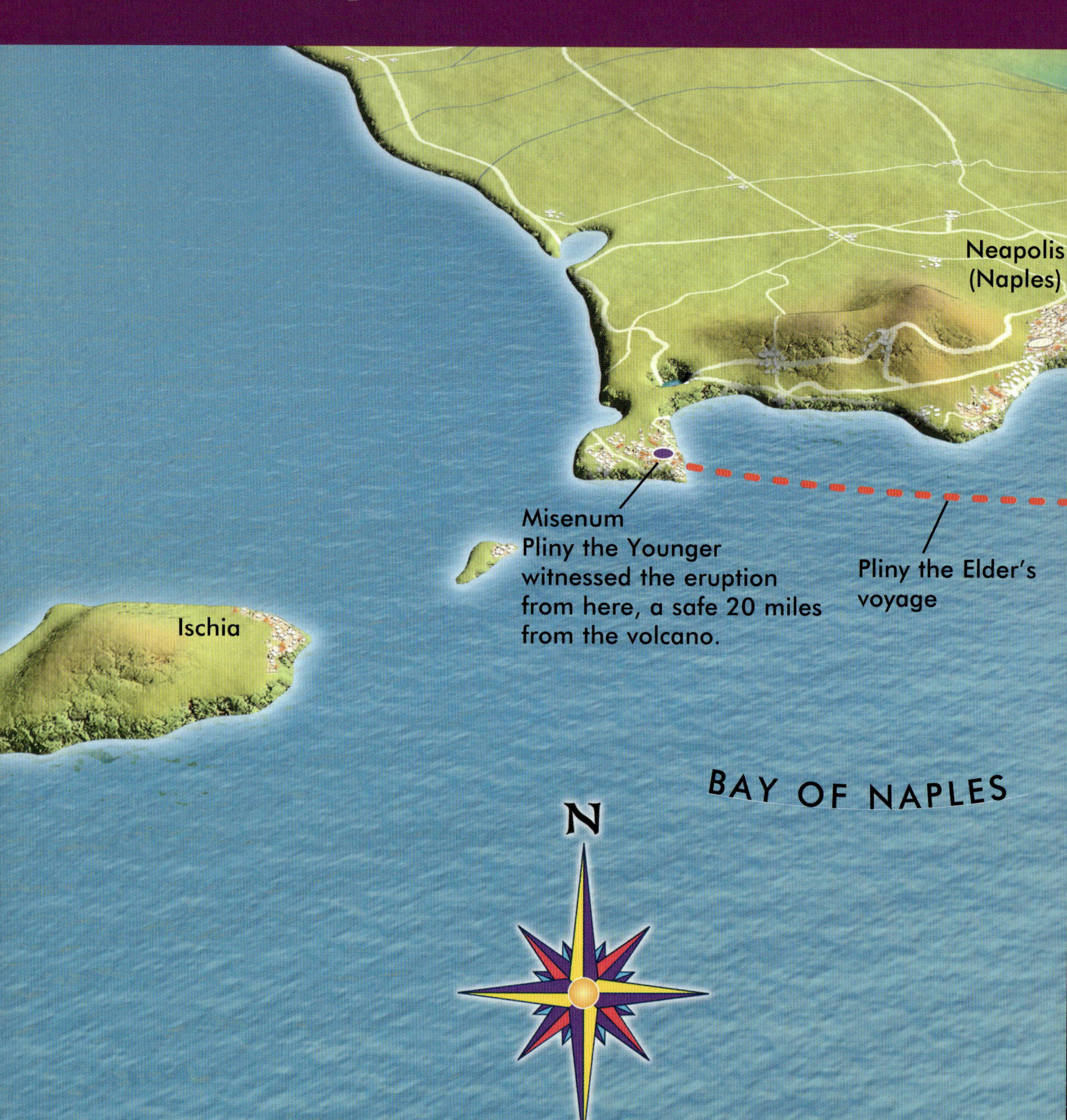

AD 79

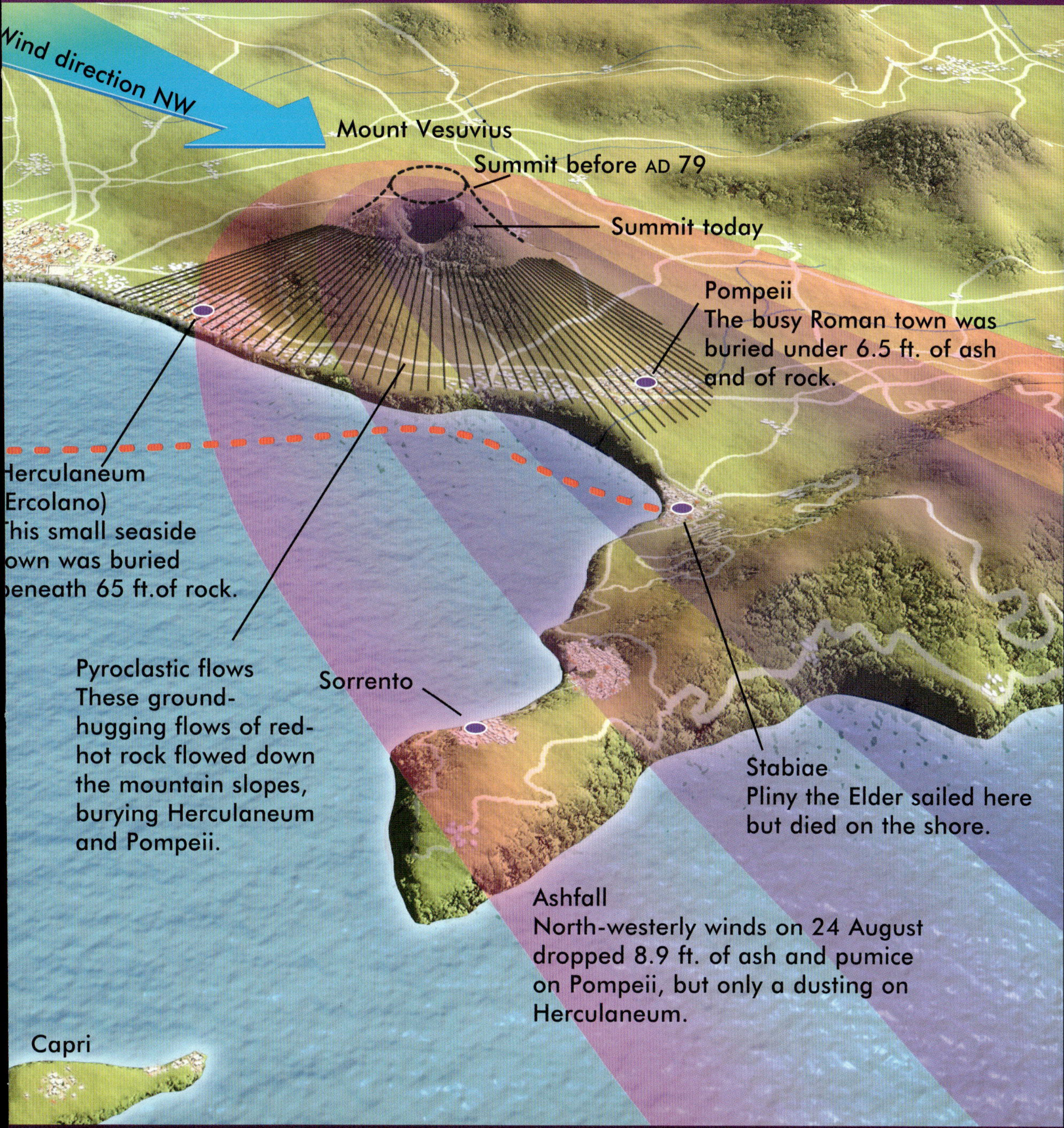

Wind direction NW

Mount Vesuvius

Summit before AD 79

Summit today

Pompeii
The busy Roman town was buried under 6.5 ft. of ash and of rock.

Herculaneum
Ercolano)
This small seaside
own was buried
peneath 65 ft.of rock.

Pyroclastic flows
These ground-hugging flows of red-hot rock flowed down the mountain slopes, burying Herculaneum and Pompeii.

Sorrento

Stabiae
Pliny the Elder sailed here but died on the shore.

Ashfall
North-westerly winds on 24 August dropped 8.9 ft. of ash and pumice on Pompeii, but only a dusting on Herculaneum.

Capri

How volcanoes form

What is a volcano?

A volcano is a hole in Earth's surface, through which hot rock, gas and steam erupt. About 62 miles below Earth's cool crust there is a hot, rocky layer called the **mantle**. The temperatures here are about 1,500° Celsius, and in some places the rock melts, forming a substance called magma. Melted magma can rise up to the surface, where it collects and bursts out through a hole in the ground. Once the magma has reached Earth's surface, it is known as lava.

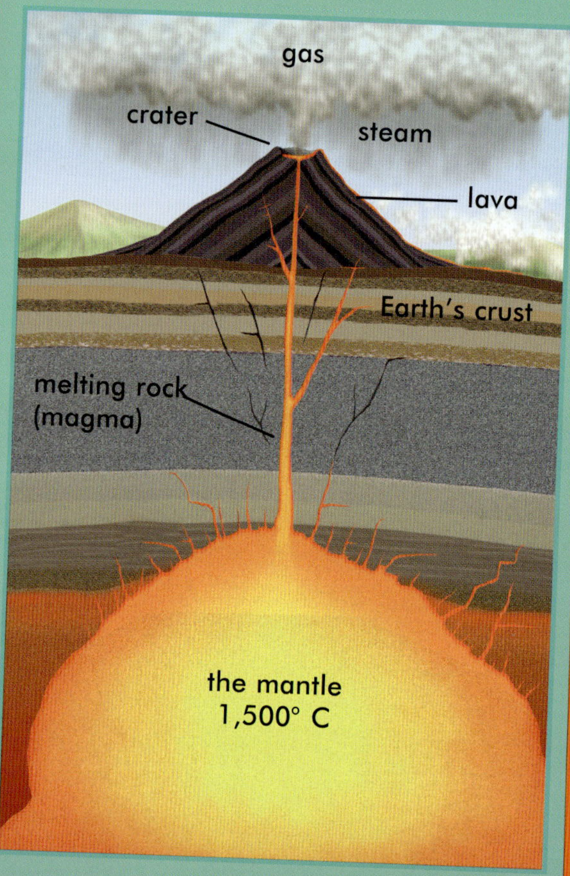

All about lava

In some volcanoes, lava is thick and sticky; in others it is thin and runny. It may ooze gently out of the ground or shoot out with ash in a violent explosion. In time, the lava cools down and sets into solid rock. The new rock may make the volcano taller, but violent explosions can destroy chunks of the volcano, making it lower as a result.

Volcano shapes

Volcanoes have different shapes, depending on the lava that formed them.

Shield volcano

Runny lava streams over the ground before setting. It forms a broad, shallow cone.

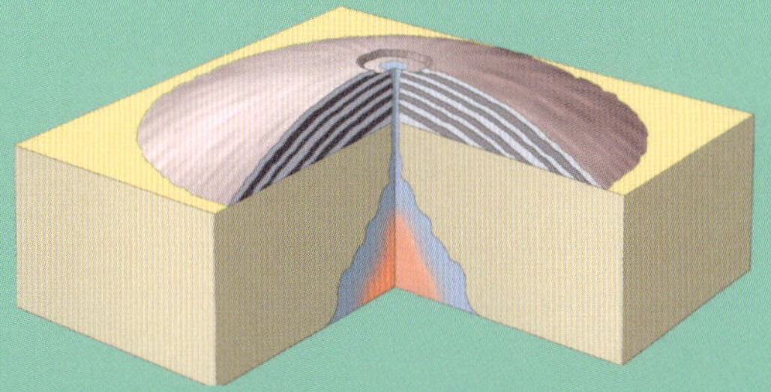

Composite volcano

Thick lava moves more slowly. It sets on top of layers of ash, to form a steep-sided cone.

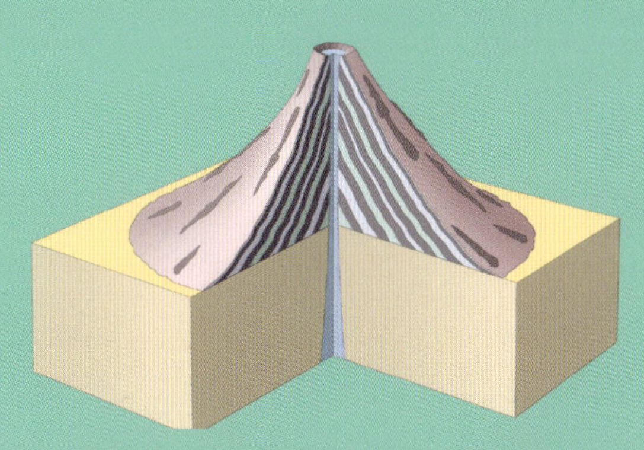

Which type of volcano is Mount Vesuvius?

Red-hot facts

- Not every eruption produces lava. The explosions of AD 79 produced hard pumice and ash.
- Most volcanoes only erupt for a brief time, separated by long periods of inactivity.
- The word "volcano" comes from Vulcan, the Roman god of fire.

Where are volcanoes found?

Volcanoes are much more common in some parts of the world than they are in others. Why is this?

Earth's plates

The outside of Earth is a hard crust, which has cracked into several huge pieces called plates. These plates move gently on the hot rock below. Most volcanoes are found near the edges of the plates. Huge forces deep inside Earth move the plates. Earthquakes and volcanic eruptions are made as the plates grind past each other.

What is a volcanic island?

Some of Earth's plates meet on the ocean floor, resulting in undersea volcanoes. The highest of these break the surface of the water, creating islands in the sea. This is how Stromboli, an island just off the coast of Italy, was formed.

Italy's volcanoes

Mount Vesuvius lies where the African and the Eurasian plates meet – under the Mediterranean region. Year by year, the African plate is moving north and forcing its way under the Eurasian plate. This is why Italy has many volcanoes, including Mount Etna on the island of Sicily, and Stromboli, a volcanic island in the Mediterranean Sea.

Red-hot fact

- The African plate is moving north at the rate of about 2 inches a year.
- There are more than 1,500 active volcanoes on Earth. About 500 of these are on land; the rest are under the sea.

Excavating

What happened to the towns of Herculaneum and Pompeii after the eruption in AD 79? The ash and rock that covered the towns created a blanket many meters deep, which preserved everything that lay beneath it.

Uncovering the town

The ancient towns were all but forgotten when, in the early 1700s, some workmen who were digging a well in Ercolano discovered the ancient remains of Herculaneum. An **excavation** of the sites began and, building by building, street by street, the ancient towns were uncovered. Fine objects such as statues and paintings were taken to museums.

Uncovering bodies

Over 2,000 people died at Pompeii. Their bodies were surrounded by ash and pumice, which set rock-hard around them. As the bodies **decayed**, they left hollows in the rock. Scientists have re-constructed the victims' body shapes by using the hollows as molds and filling them with wet plaster and then digging out the hard plaster casts. These reveal people in great distress at the very moment they died.

1 Hollow in rock made by decayed body. Wet plaster is poured into the hollow.

2 Solid plaster cast of body. The plaster sets solid and can be dug out of the rock.

A plaster cast of a victim of Pompeii.

The ruins of Pompeii have been so well preserved that from the air they could be mistaken for a modern town.

The excavation of Pompeii continues today. One third of the town still remains buried for future generations to explore.

A living Roman town

Although the destruction of Pompeii was a tragedy for its people, it has led to the preservation of an ancient Roman town. Walking around the streets today is like opening a window on life nearly 2,000 years ago.

This Pompeii bakery had four large grindstones. Mules were used to turn the stones, which ground the grain into flour.

This is one of the town's public baths. Baths were busy places where Romans bathed, exercised, met friends and played games.

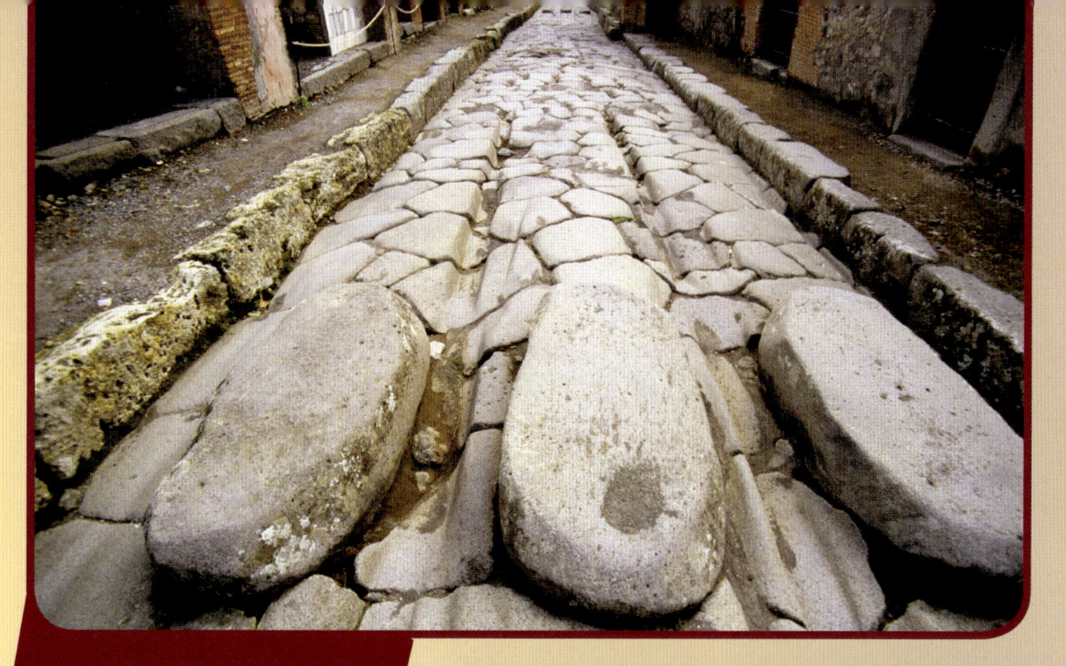

The town's streets were paved with stone. In some places they were worn into ruts by cartwheels. The large stepping stones designed for pedestrians were useful on a rainy day!

Rich Romans lived in beautiful, airy villas with fountains in the gardens, murals on the walls and decorative mosaics on the floor.

A history of

Mount Vesuvius has erupted many times since AD 79. Some of the eruptions have been small explosions with thick, slow-flowing lava. Others have been violent explosions in which hard materials like ash and pumice have been blasted into the air. Each eruption changed the mountain's shape and height.

Vesuvian eruptions (since AD 0):

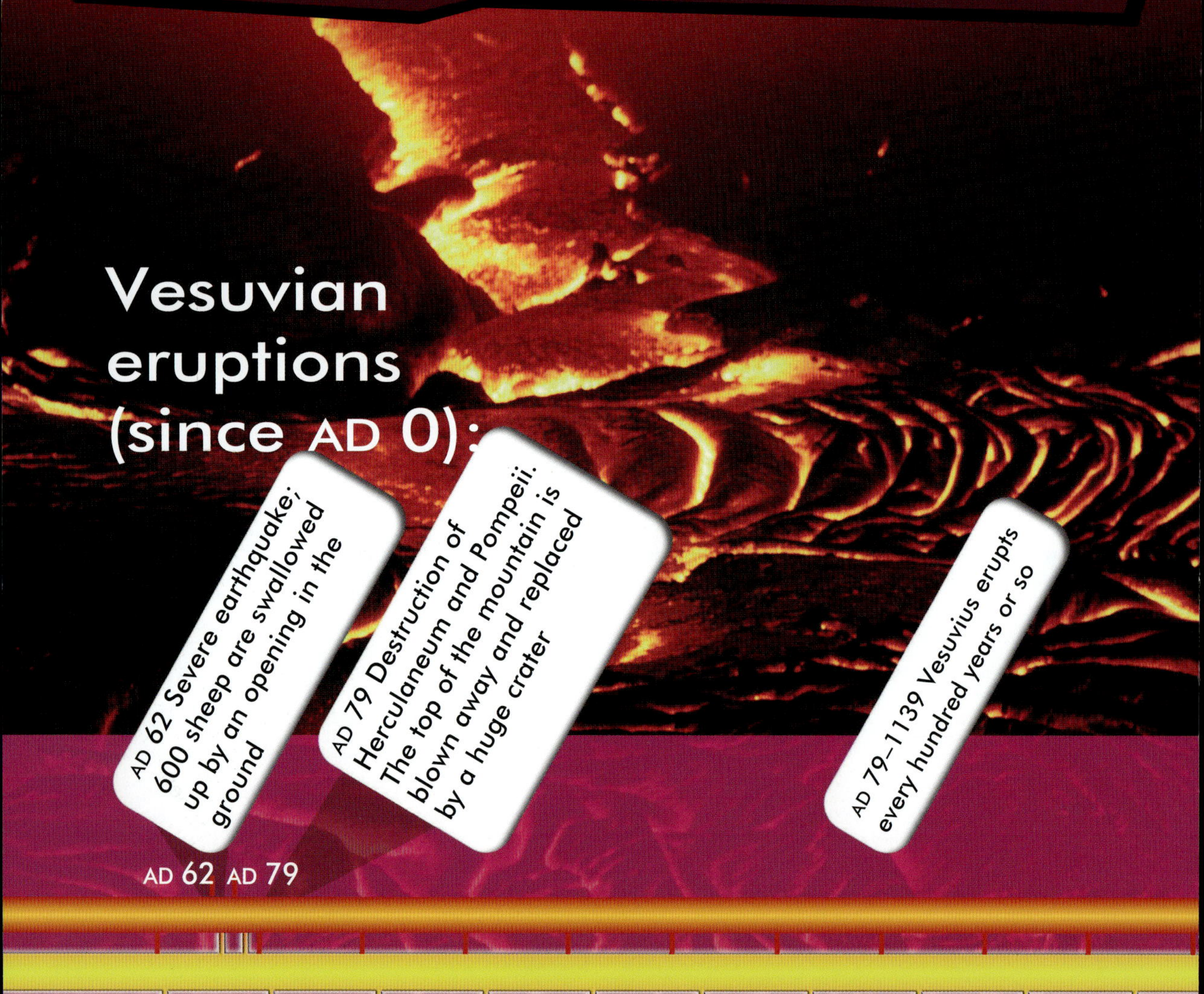
AD 62 Severe earthquake; 600 sheep are swallowed up by an opening in the ground
AD 79 Destruction of Herculaneum and Pompeii. The top of the mountain is blown away and replaced by a huge crater
AD 79–1139 Vesuvius erupts every hundred years or so
AD 62 AD 79
AD 0
100
200
300
400
500
600
800
900

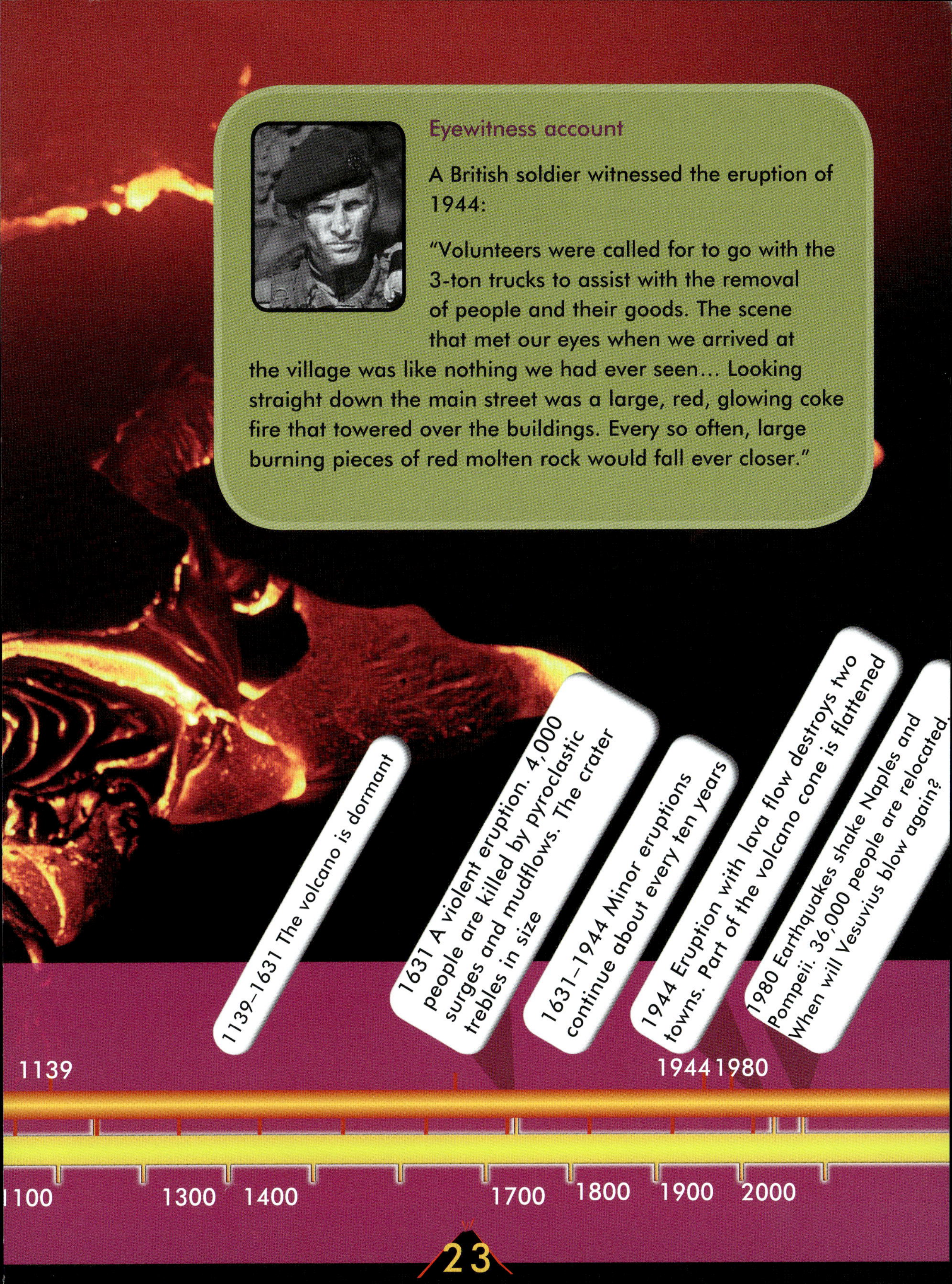

Eyewitness account

A British soldier witnessed the eruption of 1944:

"Volunteers were called for to go with the 3-ton trucks to assist with the removal of people and their goods. The scene that met our eyes when we arrived at the village was like nothing we had ever seen... Looking straight down the main street was a large, red, glowing coke fire that towered over the buildings. Every so often, large burning pieces of red molten rock would fall ever closer."

1139–1631 The volcano is dormant
1631 A violent eruption. 4,000 people are killed by pyroclastic surges and mudflows. The crater trebles in size
1631–1944 Minor eruptions continue about every ten years
1944 Eruption with lava flow destroys two towns. Part of the volcano cone is flattened
1980 Earthquakes shake Naples and Pompeii. 36,000 people are relocated. When will Vesuvius blow again?
1139
1944
1980
1100
1300
1400
1700
1800
1900
2000

Visiting Vesuvius

Mount Vesuvius is the world's most visited volcano. For about 300 years, travellers have flocked to the Bay of Naples to visit the museums and the excavations, and to climb up to the summit of Vesuvius and look into its steaming crater.

In 1800

The first travellers to Vesuvius hired a carriage to the foot of the mountain and then rode uphill on donkeys. When the slopes became too steep, the tourists continued up the mountain on foot. The ladies' dresses were clearly unsuitable for such a difficult hike!

In 1900

By the 1900s, tourists arrived at the mountain by train. Then they boarded a special mountain railway before taking a cable car to the crater rim. The railway and cable car were abandoned after repeated eruptions damaged them and they became too expensive to repair.

In 2000

Today's tourists take a train from Naples to Ercolano station and then catch a bus up the mountain. Leaving the bus park, they find a path and complete the final 328 yards on foot.

Advice to visitors to Vesuvius:

"Everyone should wear their worst clothes; boots are ruined by the sharp lava, and colored dresses are stained by the fumes of the sulphur."

From a guidebook, 1883.

Vesuvius today

Mount Vesuvius has at least a 2000-year-old history of eruptions, and yet millions of people live close by. Why do people choose to live somewhere so dangerous?

The benefits

Active volcanoes have many benefits. The ash from eruptions is full of minerals, which makes a fertile soil. Farmers who cultivate the slopes produce bumper harvests of olives and grapes and other fruit and vegetables. This is important for the whole region. Also, like many volcanic regions around the world, the area around Vesuvius is very beautiful, which attracts people to live and work there.

Then and now

In AD 79

Historians have found hundreds of amphoras at Pompeii. These tall, **terracotta** jars were used to store wine and olive oil.

Today

Two of the region's most important products are olive oil and wine.

The risks

At the moment the volcano is dormant but scientists feel sure that one day soon Vesuvius will erupt again. This is a terrifying prospect. If there were to be an eruption with pyroclastic flow, similar to the one in AD 79, there could be a tragic loss of life along with the destruction of towns, industries and farms. How can this be prevented?

Red-hot fact

- Mount Vesuvius is one of the most densely populated volcanoes on Earth with up to 30,000 people per square mile.

Volcanologists at work

Scientists who study volcanoes are known as volcanologists. They watch, measure and record volcanoes and try to predict when they will next erupt. They use electronic instruments to collect information from volcanoes. They use **seismometers** to **monitor** shaking of the ground, and tiltmeters to measure movements of the ground, caused by the magma as it rises towards the surface. Volcanologists also monitor gases inside the crater and the temperature of the ground. As well as interpreting all this scientific data, volcanologists also study past eruptions. This helps them to understand the risks an area faces in the future.

What do seismometers do?

Seismometers monitor shaking movements of the Earth – both vertical (up and down) and horizontal (from side to side). The instruments are placed on and around a volcano, and their **data** are beamed to a laboratory and recorded on computer.

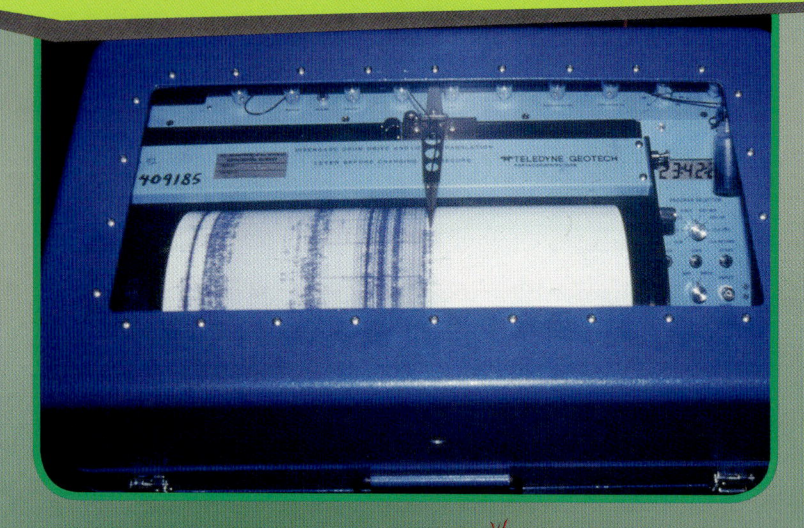

A volcanologist's equipment

Volcanologists often work on the mountain slopes. Usually, they wear their everyday clothes, but if they are working near an eruption they may need to wear a protective suit and other special equipment.

Head covering keeps out dust and gas
edc PROTECTION
Heat-resistant gloves
Suit with metal coating reflects the heat

- Volcanic gases are very smelly – like rotten cabbages and eggs! Some are so acidic they can eat through clothes.

Then and now

In 1841

Early volcanologists worked in an observatory on Mount Vesuvius – the first volcano observatory in the world.

Today

Modern volcanologists work in laboratories in Ercolano.

Escape from Vesuvius?

Predicting when Vesuvius will next erupt is very tricky. It could be many years away, but it might be much sooner. What would happen if the warning went out that Vesuvius was about to blow?

The evacuation plan

The most dangerous area in an eruption lies within four miles of the volcano. In Naples this area is known as the 'Red Zone', and is home to 600,000 people. The government plans to provide trains to evacuate everyone, but this would take a whole week to complete. Would there be enough time?

Problems with the plan

There are problems with the evacuation plan. Up to three million people might be at risk in an eruption, and earthquakes might damage the railways and roads. There is a huge need for public education, with drills to practice early-warning systems and evacuation routes. Volcanologists and planners have a difficult job, but the region and its people must be well prepared if they are to avoid the tragedy of the past.

Then and now

In AD 79
Population of Naples and surrounding area: 100,000
Today
Population of Naples and surrounding area: 3,000,000

Volcanic eruptions in films

There have been several disaster movies about volcanoes. In films, events happen very quickly. A scientist investigates a volcano with low activity, and before long people are running for their lives. In reality, monitoring a volcano and warning the public is a long, slow, careful process.

Red-hot facts

- In 1983 Naples was hit by a powerful earthquake: half a million people tried to flee in their cars, blocking streets and emergency services.
- There are 18 towns in the Red Zone. These could all be destroyed within the first 15 minutes of a serious eruption.
- A law passed in 2003 offers Red Zone residents $31,000 if they move outside the danger zone. About 2,700 people have already agreed.

What are the views of

"I was born here and I plan to die here. I hope Vesuvius doesn't kill me but if he does, that's my lot."

Vincenzo Berna, 68, a retired sailor living in the nearby town of Torre Del Greco.

"I am not afraid, I belong to this hill. Life is beautiful here. I wouldn't leave even if you paid me."

Antonio Battaglia, 75, who has lived all his life in San Sebastiano al Vesuvio, in the shadow of the volcano.

Glossary

active volcano – a volcano that erupts often

crater – the saucer-like top of a volcano

data – facts or information

decayed – rotted away

earthquake – a sudden shaking of Earth's surface that can cause great damage

excavate – to uncover something by digging

fumes – heavy, strong-smelling gases given off by a volcano

mantle – the layer of Earth which lies just below the surface. It is made up of part-melted rock

monitor – to check something regularly

pumice – a very light, silver-grey rock that is made when lava cools and becomes solid

relocate – to move home from one place to another

seismometer – a machine that monitors shaking of the ground

suburbs – an area on the edge of a city, which is home to many people

summit – the highest point of a mountain

terracotta – a type of pottery made from baked clay

tremor – a slight shaking of Earth

Index